BASIC BIOLOGY

Adaptation and Survival

DENISE WALKER

A⁺
Smart Apple Media

First published in 2006 by Evans Brothers Ltd.
2A Portman Mansions, Chiltern Street,
London W1U 6NR

This edition published under license from
Evans Brothers Ltd. All rights reserved.
Copyright © 2006 Evans Brothers Ltd.

Series editor: Harriet Brown, Editor: Katie
Harker, Design: Simon Morse, Illustrations:
Ian Thompson

Published in the United States by
Smart Apple Media
2140 Howard Drive West, North Mankato,
Minnesota 56003

U.S. publication copyright © 2007
Smart Apple Media
International copyright reserved in all
countries. No part of this book may be
reproduced in any form without written
permission from the publisher.
Printed in China

Library of Congress Cataloging-in-
Publication Data

Walker, Denise.
Adaptation and survival / by Denise Walker.
p. cm. — (Basic biology)
Includes index.
ISBN-13: 978-1-58340-992-3
1. Adaptation (Biology). I. Title.

QH546.W27 2006
578.4—dc22   2006002524

9 8 7 6 5 4 3 2 1

Contents

Introduction

The world around us is constantly changing. And as it does so, the varied species that inhabit planet Earth also have to change if they are going to survive.

This book takes you on a journey to discover more about the wonderful world of adaptation and survival. Discover the reasons behind environmental change. Learn about the ways in which animals and plants depend on each other, and look at some different examples of adaptation in action. Find out about famous scientists, such as Charles Darwin and Jean Baptiste Lamarck. Learn how they used their skills to try to explain the varied species that inhabit our planet.

This book also contains feature boxes that will help you unravel more about the mysteries of adaptation and survival. Test yourself on what you have learned so far, investigate some of the concepts discussed, find out more key facts, and discover some of the scientific findings of the past and how these might be used in the future.

Adaptation is a natural part of life on planet Earth. Now you can understand more about the reasons behind our changing world.

DID YOU KNOW?

▶ Look for these boxes—they contain surprising and fascinating facts about adaptation and survival.

TEST YOURSELF

▶ Use these boxes to see how much you've learned. Try to answer the questions without looking at the book, but take a look if you are really stuck.

INVESTIGATE

▶ These boxes contain experiments you can carry out at home. The equipment you will need is usually cheap and easy to find.

TIME TRAVEL

▶ These boxes describe scientific discoveries from the past and fascinating developments that pave the way for the advance of science in the future.

ANSWERS

At the end of this book, on page 46, you will find the answers to the questions from the "Test yourself" and "Investigate" boxes.

GLOSSARY

Words highlighted in **bold** are described in detail in the glossary on pages 46 and 47.

Why species change

There are millions of different **species** of animals and plants on Earth's surface, all living successfully side by side. This great diversity has been caused by **evolution**—a process through which species change to suit their environment. Evolution can take place over millions of years, but sometimes it happens much faster.

Evolution occurs when species are forced to **adapt** so that they can survive variations in their natural **habitat**. Adaptation may be drastic or quite subtle, but in most cases, it is a slow process—we never physically see a species change before our eyes, but we are likely to observe the signs that adaptation is occurring over a period of time.

▲ Cacti have adapted to survive the dry conditions and extreme temperatures of the desert.

WHY DO CACTI LIVE IN THE DESERT?

Deserts are very hot during the day and especially cold at night. There is very little rainfall, and the sandy soil does not support the roots of plants that grow there. Cacti have adapted to thrive in these extreme conditions:

> A cactus does not lose water from its surface like other plants. Its tough, leathery skin is covered in a layer of wax that keeps water inside and reflects the light of the sun.
>
> The spines of a cactus obstruct the wind so that the plant is surrounded by very still air, reducing evaporation from the surface. On cold nights, dew falls on the spines and is drawn toward the cactus to be absorbed.
>
> A cactus stores water in its thick stem.
>
> A cactus grows very long roots. These search for underground water and also give the cactus support to prevent it from falling over in the sandy soil.

These changes are a result of millions of years of gradual adaptation caused by **genetic** changes in generations of the cacti species. Those cacti with the smallest leaves, thickest stems, and deepest roots have survived best in the desert and have been able to **reproduce**.

TIME TRAVEL: DISCOVERIES OF THE PAST

▶ English biologist Charles Darwin (1809–1882) was the first to propose that **natural selection** leads to evolutionary change. His theory of natural selection suggested that individuals who are particularly suited to an environment survive and reproduce, giving rise to another generation of individuals with these advantageous genes. Darwin also suggested that although generations of species are similar, occasionally their genetic material undergoes a natural change that makes them more compatible to their environment. Over thousands of generations, these changes accumulate to enable a species to adapt or even to create a new species altogether.

COMPETITION

Sometimes, species have to compete with each other for space and scarce resources, such as food. In the late 1800s, European brown trout were introduced as a game fish in America, but the fish are now known to have caused a decline in some native American fish species, such as the golden trout of California. The brown trout are also known to eat frogs, birds, mice, and other small mammals.

Competition occurs in the plant world, too. In 1984, a seaweed known as *Caulerpa taxifolia* was first observed off the coast of Monaco in the Mediterranean Sea. By 1997, the weed had covered around 20 square miles (50 sq km). It is thought that the seaweed may have **migrated** from the Red Sea (through the Suez Canal) or was accidentally introduced to the sea from an aquarium. The seaweed is **toxic** to **predators** and is growing so fast that it has become a major risk to natural marine habitats.

VISIBLE CHANGES

At the beginning of the industrial revolution, a light-colored moth, called the peppered moth, was commonly seen in the United Kingdom (UK). But in the mid-1800s, a new, darker-colored moth began to be seen. At the time, coal-burning industries produced soot that caused the landscape to become blackened. Scientists believe that the peppered moth turned darker so that it remained **camouflaged** against the landscape and was less likely to be eaten by predators, such as birds. Today, however, the dark moths are declining, as the use of cleaner fuels has changed the landscape back again.

Antibiotic **resistance** in bacteria is another example of adaptation that we can observe over time. When a person is unwell, antibiotic medicines kill the harmful bacteria. However, some of the bacteria are strong and can resist the drugs. While other bacteria are killed, these stronger bacteria multiply rapidly and soon become the dominant **strain**. Over time, the antibiotics will not be effective at treating the illness.

▲ The peppered moth is light in color (left), but in industrial areas, the species has adapted to become darker (right).

TEST YOURSELF

▶ Explain how the following species may have adapted to suit their current environment.

(1) Camels that live in the desert and go for up to three months without a drink.

(2) Migratory birds that fly south for the winter.

Adaptation to environmental change can take millions of years. Environments can change suddenly (due to flooding from bad weather, for example), gradually (the effects of **global warming**), or much slower (the movement of rocks that form mountains). In our ever-changing world, species adapt to suit their surroundings. They also make daily changes to keep in optimum condition.

ADAPTATION IN PLANTS

Plants occupy most parts of our planet—we find plants on mountains, in water, and in desert sands. Like animals, plants are constantly searching for food. However, plants also need sunlight so that they can make their food through the process of **photosynthesis**. In a thick forest, only the tallest trees are able to reach toward the sunlight, making it impossible for smaller shrubs to survive.

The search for light is one of the most significant pressures that plants experience, and they have adapted to manage light levels in a number of different ways. When houseplants are left in a sunny window, they turn their leaves toward the sunlight during the day. This is because the upper surface of the leaf has adapted to capture sunlight, and the plant tilts its leaves to help this process.

▲ Kelp are aquatic plants that grow well in clear, shallow oceans where the water is less than 68 °F (20 °C).

Most plants do not like very hot and dry conditions because they lose a lot of water and are in danger of wilting. Plants in tropical and warm environments open their leaf pores during the night when the temperatures are lower. This minimizes water loss but allows the plants the opportunity to take in the carbon dioxide they need for photosynthesis.

◀ Sunflowers are one of the fastest-growing plants. They can grow up to 10 feet (3 m) in 6 months. Sunflowers need lots of sunlight to grow. During the day, the leaves and flower heads tilt toward the sun.

UNUSUAL CONDITIONS

Some plants live in more unusual environments and have adapted to face the particular challenges of their location. For example, the Venus flytrap lives in boggy soil that lacks the vital **nutrients** most plants need to survive. The plant has adapted so that it can get these nutrients from another source. Instead of using the soil, the plant captures small insects, such as flies, when they land on the plant. These insects are "digested" by enzymes so that the nutrients can be absorbed.

▲ The Venus flytrap is an unusual plant because it captures and "digests" small insects as a source of nutrients.

ADAPTATION IN ANIMALS

Animals have also adapted their behavior to fit a daily cycle. Many animals are **nocturnal,** and they come out to feed at night. This can help them to avoid hungry predators that roam during the day. Eating at night also gives an animal access to food sources that are untouched by sleeping animals.

The biggest challenge for a nocturnal animal is the dark. Most nocturnal animals have large eyes and pupils to increase the amount of light that

enters their eyes. The aye-aye, for example, lives in the rain forests of Madagascar and sleeps during the day in a tree nest. The aye-aye has large eyes to see in the dark, black hair for camouflage, and big ears to hear sounds at a distance.

▲ The aye-aye is an unusual-looking animal, but its characteristics enable it to be active at night.

INVESTIGATE

▶How have the following adaptations helped these creatures to become nocturnal?

(1) A loris
- Enormous eyes
- Big ears
- Grasping fingers

(2) A skunk
- Excretion of a very smelly spray
- Black and white coloring

Adaptation also occurs on an annual basis as conditions change throughout the year. Animals and plants have successfully adapted to cope with the varying temperatures and food resources that are found during the changing seasons.

THE PLANT CYCLE

SPRING

During the spring, many plants begin growing as part of their annual cycle. They have remained **dormant** (inactive) during the winter, but now green leaves and shoots begin to appear. As the days become longer and warmer, the sunlight maximizes the length of time that plants can carry out photosynthesis. This means that the plants have more energy to grow.

SUMMER

Summer is the time for maximum flower production. The warm days bring more insect life, too. Some plants produce colorful and fragrant flowers to encourage insects to carry the pollen to other flowers. This helps to **fertilize** other flowers, and they later reproduce. Flowering plants open their petals during the day to allow for **pollination** and close them at night for protection.

▲ Flowers bloom during the summer to aid the process of pollination so that plants can reproduce.

FALL

As the days become cooler and shorter, some leaves and flowers begin to die and fall off. Parts of the flowers turn to seeds, and these are dispersed by the wind or are embedded in fruits that are eaten by animals and distributed elsewhere. The seeds lie dormant until the following spring, when there are increased light levels to aid photosynthesis and growth.

WINTER

During the winter, plants remain dormant while they await better weather and more sunlight during the day. By losing their leaves, they save energy because they don't need to grow very much during this period. Very little appears to change during these months.

▼ Deciduous trees shed their leaves to save energy during the cold winter months.

THE ANIMAL CYCLE

BREEDING CYCLES

Although humans tend to reproduce at any time of the year, some animals, such as sheep, have adapted to produce their offspring during the spring, when the weather is more favorable for survival. Sheep and cows tend to produce reproductive **hormones** only during this time of year. However, because farmers get a good price for lambs or calves born out of season, they use hormone treatments or **selective breeding** so that a herd actively reproduces throughout the year.

MIGRATION

Many animals avoid the harsh winter conditions by moving to another area of the country, or even the world, where food is more readily available. This movement is called migration.

Every year in Tanzania's Serengeti National Park, approximately 1 million wildebeests, 100,000 zebras, and 300,000 Thomson's gazelles travel more than 620 miles (1,000 km) to escape the effects of low rainfall and scarce food resources. They risk great danger from predators along the way, and many lose their lives. However, their chances of survival are greater if they move than if they remain with limited food sources.

Animals also migrate to improve their chances of reproduction. Salmon swim great distances upstream to return to breeding grounds where conditions are more favorable. Loggerhead sea turtles also return to the beach of their birth in order to lay eggs. They often have to navigate thousands of miles through the open sea. Monarch butterflies spend the winter in the warm climate of Mexico, but in the spring, millions of monarch butterflies take flight and travel north to the

▲ Chum salmon migrate upstream in Alaska to reach the freshwater streams of their birth, where they will lay their eggs. They have to leap over numerous waterfalls along the way.

United States or Canada. They survive for only six months but manage to reproduce, and their offspring make the journey back to Mexico again. Scientists believe that some of these migratory animals have adapted to have magnetic material in their brains that lines up with Earth's **magnetic field** to aid navigation. Other animals are believed to use an acute sense of smell to find their way.

HIBERNATION

Some animals sleep through the winter months. This is known as **hibernation**. Hibernation helps animals to survive cold weather and a lack of food supplies. During hibernation, life processes such as breathing and feeding slow down or stop completely. When humans are inactive for long periods of time, their muscles waste away, and there can be a temporary loss of movement. Scientists believe that a type of shivering helps hibernating animals to preserve their muscle strength.

Predators and prey

The world in which we live is constantly changing as individual species adapt and environments alter. Sometimes, if adaptation does not occur quickly enough, species have to compete for the scarce resources that are available. Some animals prey on each other as a source of food.

PREDATION

Predators kill other species for their food, and they often use clever strategies to catch their **prey**. They may work in a group to increase their chances of a successful hunt. Predators may also choose vulnerable animals (such as the old, young, sick, or injured) as easy targets. Some predators migrate to areas where there are a lot of prey. This prevents them from having to depend on just one species for their food.

Predators who hunt singly among mammals are usually large, powerful animals that can easily overcome their prey. Predators of this type usually have the following characteristics:

Excellent eyesight and hearing for spotting prey.

Powerful muscles so that they can run or swim fast to catch their prey.

Camouflaged coat so that they can get close to prey without being seen.

▲ The leopard's spotted coat provides almost perfect camouflage in grass or thick brush.

◄ Wolves hunt in packs to increase their chances of a successful catch. A pack of wolves can overcome animals that are much larger and stronger than themselves. The pack also defends and guards the territory where it hunts and lives from other intruding wolves.

PREY

Prey animals can reproduce and pass on their genes only if they avoid predation. They adopt a number of strategies to ensure the survival of their species. Some prey animals:

Reproduce frequently so that when some are caught, others can ensure the survival of the species.

Run very fast. For example, rabbits can run faster than foxes.

Stay in large groups so that they can all watch for predators while they feed.

▲ Deer live in herds to minimize their chances of being attacked by a predator.

Adapt so that they taste horrible to predators. For example, monarch butterflies taste bad and can be poisonous to a bird if it eats too many of them.

Camouflage themselves.

Adapt to look like other species that avoid predators. For example, hoverflies look just like wasps, but they do not sting.

Adapt color patterns that deter their predators. For example, ladybugs and some moths and butterflies are brightly colored to suggest that they are poisonous.

TEST YOURSELF

▶ In what ways is a tiger adapted to prey on other animals?
▶ In what ways is a hare adapted to avoid predators?

PREDATOR-PREY NUMBERS

When predators catch their prey, they impact the number of prey species in an environment. If predators catch more prey than are reproduced, the prey species will eventually become **extinct**. The link between the numbers of predators and the numbers of prey is summarized by the graph below:

RELATIONSHIP BETWEEN PREY AND PREDATORS

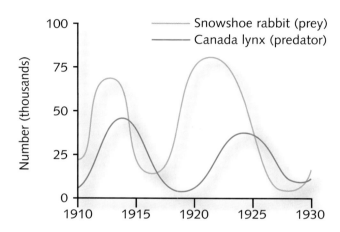

This graph shows the relationship between the numbers of snowshoe hares and Canada lynx between 1890 and 1930. Notice how the lines change in a cyclical pattern. If prey have plenty of food, their numbers increase, as they feed well and reproduce. This means that predators now have a plentiful supply of food and in turn grow and reproduce. As the number of predators increases, more prey are eaten, and the numbers begin to fall again. This results in less food for the predators, so their numbers also start to fall. When there are fewer predators, prey numbers increase once more, and the cycle continues over and over again.

The predator-prey cycle is also affected by natural fluctuations in numbers as a result of breeding seasons because some animals breed only at certain times of the year. In addition, there can be high numbers of deaths within populations if harsh environmental conditions occur, such as a drought.

HUMAN HUNTERS

Since the dawn of time, men and women have been hunter-gatherers, eating wild plants, nuts, and fruits, and hunting animals for their survival. As an intelligent species, they have adapted both physically and behaviorally to hunt effectively for their food.

▲ Early man used tools made of wood and stone to hunt for food.

Around five million years ago, our ancestors adopted a bipedal (upright) posture. Walking on two legs made movement easier and also left hands free for carrying tools for hunting prey.

With this change in posture came changes in the circulatory system and the spinal column. Today, the human brain is about three and a half times bigger and more complex than the brain of our early ancestors. Since intelligence and the ability to communicate were necessary for successful hunting expeditions, the more intelligent survived and passed their genes to the next generation.

HUNTING TODAY

Some humans still hunt for their food, and with the development of technology and skills, this has become easier. In ancient times, humans used spears to kill their prey. Later, arrows were introduced so that hunters could keep at a distance from their prey. Now, weapons are more sophisticated. Animals can be seen from long distances, and the prey may not see its predator at all.

FARMING

Today, instead of hunting, it is more usual for humans to use farming techniques to make their food readily available. Animals such as cattle, pigs, and chickens are brought up on farms and killed for meat that can be sold and eaten. Animals are bred so that their number meets demand. This stops species from becoming in danger of extinction.

THE EFFECTS OF HUNTING

As fishing techniques have advanced, many marine species are now threatened with extinction. More fish are hunted than can be replaced naturally, and many marine mammals get caught on fishing hooks or in fishing nets.

Whales are hunted for their meat, oil, and whalebone. In 1986, a decline in numbers led the International Whaling Commission to propose a ban on whale hunting. It is difficult to record the number of whales traveling across the oceans, but numbers have dropped dramatically.

In Africa, elephants are **poached** for their ivory, and rhinos for their horn. These products are so valuable that it is difficult for hunters to resist. In eastern Africa, the black rhino is on the verge of extinction, and elephants are also threatened. Wild animals are also hunted because their meat can be sold commercially. Today, one in three **primate** species faces extinction due to hunting.

Food chains

Plants are important to our planet because they use photosynthesis to provide animals with the energy they need to grow and carry out life processes. When plants are eaten, their energy is transferred to another species. We call this a food chain.

There are three basic levels of a food chain:

Level 1—Plants are called producers because they have the ability to convert the sun's energy into chemicals that can be used to power life processes. Plants capture about 0.01 percent of the energy of the sun's rays every day and are the basis of the whole food chain.

▲ Trees are producers. They use light energy from the sun to produce food from carbon dioxide and water.

Level 2—Animals on the second level of a food chain are called primary consumers. We call them herbivores because they eat only plants for their energy. For example, deer eat shrub leaves, rabbits eat carrots, and worms eat leaf litter. Primary consumers need to eat a lot of plants to meet their daily energy requirements. They eat approximately 10 percent of the available plants and gain about 0.001 percent of the sun's original energy.

Level 3—Animals on the third level are called secondary consumers. They eat primary consumers for their energy. For example, lions eat deer, foxes eat rabbits, and birds eat worms. Secondary consumers are good predators, but there are fewer of them because their energy sources are limited. Secondary consumers usually eat meat, and we call them carnivores (if secondary consumers eat mainly insects, we call them insectivores). Secondary consumers gain about 0.0001 percent of the sun's original energy.

Further levels of the food chain are possible, but the longer a food chain is, the less energy is left for the species at the top of the chain.

▲ Lions are secondary consumers. These lionesses are eating a zebra that has been hunted.

REPRESENTING FOOD CHAINS

We can represent food chains using arrows like this:

Plants 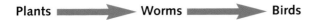 **Worms** **Birds**

The arrow represents which species eats which. That is:

Plants   (eaten by)   **Worms**   (eaten by)   **Birds**

HUMANS AS CARNIVORES

Humans are at the top of their food chain because other animals don't naturally eat them. Humans are called omnivores because they can eat a variety of foods, such as plants and animals. Humans are not the largest of all species, but because they have superior brains, they often eat animals that are larger than themselves. This is because they have used their intelligence to develop farming and hunting techniques.

A MORE EFFICIENT FOOD CHAIN

Vegetarians and vegans are mostly herbivores (primary consumers). They make the human food chain shorter, so there is less energy lost at each level. Nonvegetarians have also developed ways of making their food chain more efficient:

▶ They use factory farming instead of free-range farming. Chickens are kept in very small areas of space, which means that they cannot move around and therefore do not use energy. Some people believe that this is cruel and prefer to buy free-range products, which allow animals to grow up in more natural conditions.

▶ They feed animals with grass—a food source that is not eaten by humans. Many animals are also fed grains and crops. When people eat these animals, they benefit from the energy of the grass, grains, and crops.

TEST YOURSELF

▶ In what ways do food chains "lose" energy?

▶ Give a definition for each of the following: herbivore, omnivore, carnivore, insectivore.

▶ Give an example of a food chain that has humans at the top.

▶ Identify the producers, primary consumers, and secondary consumers in the picture below.

Mad Cow Disease

In 1986, bovine spongiform encephalopathy (BSE), or "mad cow disease," was first identified in cattle in Britain. Since then, more than 160,000 BSE-infected cows have died or had to be killed. Meat exports caused this brain disease to spread to cattle in Europe, Japan, and America, although strict testing has helped to contain it. BSE is very rare in humans, but in 1986, the first human form of BSE (called variant CJD) was identified. This disease affects the human brain and may be caused by eating BSE-infected beef, although there is no definite link. In 1989, the British government began a policy of disposing of any infected animals to avoid passing BSE down the food chain. In 1993, 52,000 cows were diagnosed with BSE. This figure has now dropped to around 398.

FOOD WEBS

Feeding relationships are complicated by the fact that conditions in an environment can change. In order to ensure survival when food sources are scarce, animals usually eat more than one thing. Diseases can also cause a species to decline in population, and this can have an effect on various food chains.

The diagram below shows the feeding relationships between different species in a marine environment. Notice how some species feed on two or more other **organisms**.

MARINE FOOD WEB

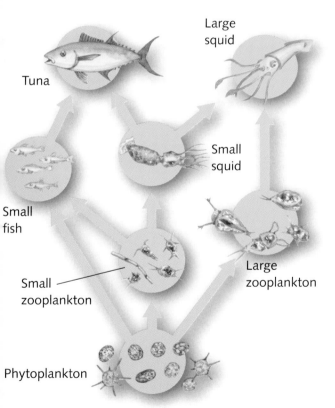

Tuna eat small squid and small fish, but what would happen if the small squid population died out? Tuna would feed on alternative food sources, but this would mean that the numbers of small fish would drop. The number of tuna

may also fall slightly if there isn't enough fish to go around. Alternatively, what would happen if the small squid and small fish all died out? The tuna would no longer have any food source, and its numbers would fall dramatically. This scenario can happen in extreme cases if species are hunted to near extinction or affected by a deadly disease. However, it can also be caused by annual changes if whole species disappear during migration or hibernation.

A CHANGING FOOD WEB

In the North Sea, the following food chain exists:

Copepods → Krill → Norway pout → Cod

Cod are being overfished in many seas (see page 14), and as cod stocks decline, fishermen move down the food chain to fish other stocks (such as Norway pout). Traditionally, fishermen have looked for fish stocks in waters nearest the coast. During the last 30 years, however, they have begun to explore farther from the shore, where oceans are deeper. This means that fish that have taken a long time to reproduce are now being harvested. The orange roughy found off the coast of New Zealand, for example, can take as long as 30 years to reach sexual maturity and reproduce. Because this fish is now being hunted, its population has fallen by about 80 percent and is currently showing no sign of recovery.

> ## INVESTIGATE
>
> ▶ Draw a food web for a forest environment and a food web for a pond environment. You will need to do some research into the food chains in these environments before you begin.

TIME TRAVEL: POLLUTION

Pollution occurs around the world because chemicals or substances are leaked into the environment. Food chains can be dramatically affected as natural habitats are destroyed or species are harmed by the effects of pollution. Some pollutants have had a particularly damaging effect on the wildlife of our planet.

DDT

In Africa, malaria is a disease that kills a million people a year (mostly children). DDT (dichloro-diphenyl-trichloroethane) is a chemical that effectively kills mosquitoes, the carriers of malaria. DDT can be sprayed on crops and in the home.

▲ **This farmer is spraying DDT powder to kill pests near his village in Namibia.**

During the 1950s and 1960s, DDT was sprayed all across southern Europe, where it eliminated malaria. It was also commonly used in Africa until the late 1970s. DDT was successful in many ways because it was fairly cheap to buy, and its effects were long lasting. The pesticide was popular in the fight against insect-borne diseases, but it was also used to protect crops. In the U.S., for example, DDT was sprayed on cotton crops to kill pests that might destroy the harvest. However, DDT also brought many problems.

WHAT HAPPENED?

Although DDT was one of the most powerful weapons in the war against malaria and crop control, it was also lethal to the environment:

(1) Flooding washed the chemical everywhere.

(2) DDT was found to kill fish and to threaten birds.

(3) Although no studies actually showed that DDT was harmful to human health, it has been widely suspected of being so.

(4) DDT entered the food chain and was found to thin the eggshells of peregrine falcon chicks. The birds accumulated DDT in their body by feeding on birds that had eaten DDT-contaminated seeds or insects. Due to extensive spraying of DDT, the eggshells often broke during incubation or failed to hatch at all. When too few chicks were born to replace the adults that died, peregrine populations declined dramatically.

DDT was banned in most countries in 1972, but by this time, the number of peregrine falcons had fallen by about 80 percent. After the ban, numbers gradually began to increase again, and although the effects were not immediate, thousands of peregrines are now found worldwide (except in the rain forests and the cold, dry Arctic regions).

▲ **Peregrine falcons were in danger of extinction in the 1970s due to the effects of DDT spraying.**

The DDT story is an important lesson because it shows us the effect of **bioaccumulation** and **biomagnification** (see page 46). Even if small amounts of pesticide leak into the environment, they can form a dose high enough to cause problems higher up the food chain. DDT destroys insects, but the animals that feed on those insects consume the pesticide, which is then passed to the next species in the food chain. Secondary carnivores, such as the peregrine falcon, consume more prey than primary carnivores, so the levels of DDT become much greater in these animals. Pesticides such as DDT are also slow to break down. If 220 pounds (100 kg) of DDT were distributed, there would still be more than 2.2 pounds (1 kg) remaining after 100 years.

GLOBAL CONCERN

A number of pesticides have now been linked to a variety of health concerns, including lowered intelligence and weakened immune systems in seals. Pesticides have become such a concern to countries that in 1998, 120 nations met to discuss a worldwide ban on such chemicals. Of the 120 nations that met to discuss the use of pesticides, northern countries (where pesticides tend to settle) generally favored an international ban. Some

southern countries, where insect populations are higher and disease is widespread, were more reluctant. Politicians in these countries tend to judge the ecological risks of using pesticides against the dangers that the spread of diseases, such as malaria, brings.

WHAT ARE THE ALTERNATIVES?

Countries are looking at alternative pest control strategies to reduce the amount of pesticides used. Alternative methods include biological controls (for example, introducing a pest's natural enemies into the area) and timing planting and harvesting to avoid peak activity by pest species.

DID YOU KNOW?

▶ About 110,000 tons (100,000 t) of old and unused toxic pesticides have been abandoned in sites around Africa and some areas of the Middle East. These chemicals were placed in drums that are beginning to leak in both rural and urban settings. This leaking can endanger human health and lead to respiratory problems and headaches. In fact, researchers have found more than 5,500 tons (5,000 t) of DDT in Ethiopian villages. It will cost about $250 million to clean up this debris.

▼ In 1998, the United Nations began negotiations to ban DDT and other toxic chemicals known to cause damage.

▶ A recent survey has shown that traces of a type of pesticide called chlordane used in North America, Europe, and Asia are now polluting some parts of the Arctic. When pesticides are sprayed on crops, they can enter the atmosphere and travel on the wind. As the chemicals reach the colder parts of the globe, they condense and fall in rain. Chlordane was banned about five years ago because the chemical was found to be carcinogenic (cancer-causing). However, Canadian scientists have found significant amounts of chlordane in the Arctic atmosphere, showing that this pesticide is proving to be a persistent chemical.

Many chemicals can cause problems for food chains. Man's worst disasters have caused gallons of oil to be spilled, with devastating effects on the surrounding environment. The use of heavy metals in industry has also brought the problem of what to do with potentially harmful waste. Toxic metals have leaked into the environment, sometimes causing irreversible effects.

▼ Oil tanker *Fair Jolly* (left) drains crude oil from the grounded Greek ship *Tasman Spirit*, which broke in two after it ran aground off the Pakistani coast in 2003.

OIL SPILLS

In July 2003, *Tasman Spirit*, a Greek oil tanker traveling to Pakistan, spilled 30,000 tons (27,200 t) of oil into the sea when it ran aground in shallow water. The area was rich in mangrove forests, which provide a safe home for sea turtles, dolphins, and porpoises, as well as those species lower in the food chain such as shrimp, crab, and lobster. Aside from the direct damage to the environment, 12,000 tons (11,000 t) of oil evaporated into the atmosphere, causing an overpowering odor.

Oil spills can have the following effects on food chains:

A thin film of oil remains on the surface of the water, preventing sufficient sunlight from reaching the plants of the sea. The producers at the bottom of the sea food chain are reduced dramatically in number.

The sand on the shore soaks up the oil, leaving no gaps for animals on the shore to occupy. This particularly affects polychaete worms. These worms are an important source of energy for other species. Other beach life may also be affected, such as clams and starfish.

Fish in the area become coated in oil that clogs up their fins and gills. The fish cannot swim effectively or take in the oxygen they need for respiration. This can affect fish such as eels and catfish, both of which are important prey for animals farther up the food chain. Damage to the insulating fur or feathers of animals and birds can cause them to sink or die from hypothermia.

Oil that reaches mangrove areas affects the young plants that are growing in these regions. One of the most important mangrove plants is *Avicennia marina*, which cannot grow in oil-filled conditions. Without its ability to grow and reproduce, the mangrove is seriously affected.

▼ When a bird becomes coated in oil, its feathers stick together and no longer provide a waterproof, buoyant, and insulating covering.

TIME TRAVEL: INTO THE FUTURE

▶ Scientists have found that certain plants are capable of absorbing toxic waste. In the future, these plants could be used to clean up the environment. The plants could do this in two ways:

(1) Absorb metals and store them in their cells. Once this work is done, the plants could then be harvested and the material buried. For example, mustard plants are known to absorb nickel and copper.

(2) Be artificially modified to absorb toxic metals and convert them into a form that is less dangerous. For example, a plant called *Arabidopsis thaliana* has been modified with an artificial gene, taken from bacteria, so that it can withstand high levels of mercury. This may open the possibility of cleaning up toxic metals through planting the right species. The use of biological organisms to clean up the environment is called bioremediation.

Other significant oil disasters at sea include that of the *Prestige* oil tanker, which spilled 70,500 tons (64,000 t) of oil off the northwest Spanish coast in 2002. The boat's hull cracked open during a severe storm, and high winds drove the oil toward the shore. In 2001, *Jessica* struck a reef off San Cristobal Island in the Galapagos Islands, spilling oil that killed a wide range of wildlife. These disasters are very costly in terms of species loss but also have serious financial implications.

METAL AND CHEMICAL SPILLS

Metals such as mercury, copper, and lead have also been leaked into the environment from metal processing and mining. These metals are poisonous and cannot be broken down, so they become more concentrated in the food chain.

In February 2000, a storage pond near a gold mine in Romania burst its banks. The water that leaked out contained heavy metals that spilled into small local rivers and eventually reached rivers near Hungary. The spill killed thousands of fish and other animals in the food chain, and poisoned local drinking water. Two years later, Hungarian fishermen's catches were still one-fifth of their original levels.

In 2005, an industrial blast in Jilin, China, spilled more than 110 tons (100 t) of benzene into the Songhua River. The river serves as the main drinking water supply to Harbin, a city of nearly 4 million people, situated 160 miles (260 km) downstream. As a 50-mile-long (80 km) slick of the toxic chemical moved downstream, Harbin's water supply had to be shut down for five days.

▲ Researchers take samples of polluted water near the U.S.-Mexico border to determine the effects of pollution.

Conservation and biodiversity

Conservation is a way of keeping species and natural habitats as stable as possible while physical factors, such as natural disasters or man-made accidents, threaten species and affect associated food chains. Humans also try to actively maintain the number of different species on our planet by creating safe habitats in which they can thrive.

PLANT CONSERVATION

Plants are constantly under threat of extinction. In 1998, a report claimed that at least one in eight plant species is in danger—threatened by persistent predators and the destruction of land through deforestation, drainage, and development.

Madagascar is a country that has unique orchids and forests that span many miles. However, as a poor country, its economy needs investment from companies that wish to extract precious metals from its forests. Cobalt and nickel reserves worth about $2.25 billion have been found in Madagascar—but forests will be lost if roads and pipelines are built to transport the materials.

During the 1990s, a number of environmental organizations warned that the genetic diversity of crops around the world was rapidly falling, causing a reduced resistance to pests and diseases. This would threaten a shortage of food for millions of people. If plant genetic variety is reduced, a disease that kills one type of plant is very likely to kill another due to the genetic similarity of the plants. For example, there is concern that breeding corn species to produce profitable harvests is causing the species to become less diverse, making corn more susceptible to bacteria, fungi, and insects.

In 1996, research by the United Nations showed that only 200 of the planet's 500,000 edible plants were cultivated. Thousands of varieties of plants have been lost, and a predicted million more will disappear unless action is taken. Modern commercial agriculture is one cause of the losses. Many species are now being crossbred to improve the taste of fruit or the size of a crop. When crossbred varieties are planted, local varieties begin to die out. One way of conserving plant species is the use of seedbanks (see page 23).

◀ Swamp forests are a source of rich vegetation, but many are under threat.

SEEDBANKS

Many countries have seedbanks that store thousands of seeds from plants from around the world. For example, the Royal Botanical Gardens at Kew, England, and the National Seeds Storage Laboratory (NSSL) in Colorado, hold seeds. The world's largest and oldest seedbanks are in St. Petersburg, Russia, and contain around 380,000 seed varieties.

The seeds are kept for conservation, but also for research and educational purposes. Seeds are kept in cold storage and can be used to regenerate natural vegetation. However, many seeds have been lost because they have not been stored in the right conditions.

ANIMAL CONSERVATION

Many animals are endangered because they have been hunted or overfished. Animals are also in danger as a direct result of human actions, such as tourism and building. Conservation initiatives have now been set up to try to address these problems.

For example, Tanzania has increased the number of an endangered turtle species along its coastline through a program of education and awareness. On Mafia Island, just off the coast of Tanzania, communities monitor and protect the turtle nests in their villages. Villagers are paid $10 for every nest they discover and protect, and activities are reported to their village monitor. Fishermen are also encouraged to release live turtles that are

caught. The project began in 2001, and the number of recorded nests has doubled from 68 to more than 150 a year. Turtle poaching has also fallen by 90 percent since the beginning of the program.

WILDLIFE CORRIDORS

Humans build houses, roads, railroads, and bridges in many rural areas, causing environments to become fragmented or even destroyed. Like humans, some animal species need to cross busy roads to travel from one part of their habitat to another. For this reason, wildlife corridors have been set up to allow the movement of animals and plants from one place to another. They may include ditches, hedgerows, rivers, or river valleys. Wildlife corridors not only link areas of habitat, but they can also provide new habitats themselves.

▲ This fence in Canada is used to guide snakes to a tunnel under the road to protect them from passing vehicles.

INVESTIGATE

▶ Imagine that you are in charge of a panel that needs to decide whether a company should be allowed to destroy a rain forest in Madagascar for precious metal extraction. What arguments would you expect to hear from both sides? What decision would you make and why?

Pyramids of numbers

Food chains and food webs show us how energy is transferred along a food chain. However, we use what we call a pyramid of numbers to tell us how many individuals are in each link of a food chain.

CONSTRUCTING A PYRAMID

We represent the numbers of species in each link of the food chain using a pyramid of numbers. Consider the food chain on page 16.

Plants ➡️ **Worms** ➡️ **Birds**

Worms must eat many plants to gain their energy requirements. If there were too few plants, the worm population would fall. Birds are larger than worms and need to consume a number of worms in order to survive. There are more plants than worms, and more worms than birds. This information can be represented as a pyramid of numbers.

A pyramid of numbers helps us to see visually how many individuals there are in each link of the food chain. The size of each rung tells us approximately how many individuals that level contains.

A PYRAMID OF NUMBERS

Each rung shows us the feeding habits of the different organisms within the food chain. Each rung is called a **trophic** level and should be drawn to a roughly comparative scale.

Consider the food chain on page 16.

INVESTIGATE

▶ Draw out pyramids of numbers using the following information.

(1) 5 seabirds; 50 mussels; 5,000 small algae

(2) 2 foxes; 15 weasels; 65 voles; 369 oak leaves

(3) 2 birds; 100 aphids; 1 oak tree

▶ What is wrong with the last pyramid?

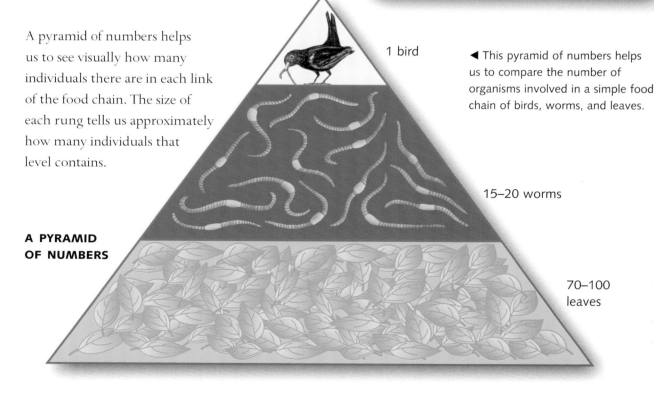

1 bird

15–20 worms

70–100 leaves

◀ This pyramid of numbers helps us to compare the number of organisms involved in a simple food chain of birds, worms, and leaves.

PROBLEMS WITH PYRAMIDS

Sometimes the number of organisms can be very large at one level and very small at the next. This would make drawing a pyramid to a comparative scale very difficult.

Sometimes pyramids of numbers can appear top-heavy. Consider the following food chain.

Rosebush ⟶ Greenfly ⟶ Ladybug ⟶ Parasite

There are many parasites feeding off of one ladybug, so the pyramid of numbers would be like that shown to the right.

Sometimes the pyramids appear to be upside-down. Consider a food chain in which one oak tree feeds many insects, that in turn are preyed on by blue titmice. This pyramid does not consider the size of the individuals at each trophic level.

Parasite

Ladybug

Greenfly

Rosebush

PYRAMIDS OF BIOMASS

If a pyramid of numbers gives an upside-down shape, a much more accurate way of representing the information is to use a pyramid of biomass. Biomass takes into account the size and weight of organisms. Scientists capture, count, and weigh a sample (the branch of a tree, for example, which can be used to estimate the tree's total weight). The pyramid changes shape completely if we consider the comparative weights of organisms at each trophic level. The pyramid of biomass shown at right indicates that:

▶ The oak tree is the producer.

▶ The caterpillars are primary consumers. They are herbivores that eat some of the available plant material (mostly from the leaves). The

caterpillars actually convert about 20 to 60 percent of the oak tree biomass into their own biomass. Some of the energy is used for living processes, such as respiration. The trophic level for the caterpillars is smaller than that of the tree—although they are greater in number, they are much smaller in size.

▶ The blue titmice are carnivores.
Some of the caterpillars are eaten. Others defend themselves by hiding. In comparison to the caterpillars, blue titmice can convert a greater amount (about 50 to 90 percent) of the biomass of their food into their own biomass.

Our pyramid of biomass has three trophic levels. Most food chains have no more than five trophic levels because when energy and biomass are lost at each trophic level, they cannot be sustained.

▼ An oak tree is very large. It feeds quite a few caterpillars, but they are small in size. A blue titmouse is slightly larger, but it feeds on a number of caterpillars.

Bird

Caterpillars

Oak tree

METHODS OF SAMPLING

We collect data for pyramids in a number of ways:

Quadrat	This wooden frame is randomly placed over a piece of land (or even the ocean floor). The organisms that fall within the square are counted. Several quadrats are recorded to give an overall picture of the environment.
Transect line	This straight line is placed through a habitat. The organisms on either side of the line are identified and counted. Transect lines show how environments change over a given distance.
Pitfall traps	These small traps can be used to catch small animals and are often used to study the movement of animals in an environment.
Pond dipping	This method is used to capture aquatic animals both near the surface of water and deeper down. Once all of the organisms have been examined and counted, they can be returned to the water.

DRAWING A PYRAMID OF BIOMASS

Scientists need to collect and weigh samples so that they can plot a pyramid of biomass. This is very time-consuming unless a small number (sample) of organisms is used to represent all of the organisms within a given environment. Random samples (of a good size) are considered a fair representation of the environment as a whole. This can be achieved by sampling several times.

▲ Nets can be used to take samples from the water when pond dipping.

SHORTFALLS OF PYRAMIDS OF BIOMASS

The pyramid of biomass is generally a much better way of representing organisms in a given environment, but it is not without its own problems. For example:

▶ Samples are taken at one period of time. They don't take into account changes in eating patterns, such as grazing on fresh grass.

▶ Biomass changes with the seasons (for example, some trees shed their leaves in the winter).

INVESTIGATE

▶ Construct a transect line six and a half feet (2 m) long from the base of a large tree outward. Every eight inches (20 cm), identify and count the number of organisms found to the left and right of the transect line. Construct a pyramid of biomass from the data you have collected.

Habitats and ecosystems

All living organisms choose a place to live based almost entirely on four factors: Is the environment safe? Does the environment offer shelter? Are there plenty of food sources? Is the environment suitable for breeding new generations? If all of these requirements are satisfied, the organism is likely to live in the area all of the time. But if only some factors are true, the organism may need to move elsewhere at particular times of the year.

If we look at the organisms that occupy some of the extreme places on Earth, we can begin to understand the attractions of their habitat. For example, the harsh conditions of the North Pole may not look very inviting, but because the area is not densely populated, the organisms that live there have very few predators.

factors include the amount of oxygen, sunlight, and water in that environment. Organisms don't depend on just each other for their food. For example, fish need underwater plants to supply the water with oxygen, some plants need insects to pollinate them, and birds depend on trees to provide suitable nesting places.

DESCRIBING AN ENVIRONMENT

Wherever an organism lives, we classify parts of its environments with some key terms:

Population	All of the members of the same species in a given area.
Community	Population of all of the different species in a given area.
Habitat	Part of an environment that offers food, shelter, and breeding grounds.
Ecosystem	All of the living and nonliving parts of an environment.

When we look more closely at a living organism's environment, we can break it down into living (biotic) parts and nonliving (abiotic) parts. Biotic factors are things such as food availability, predators, and mates. Abiotic

▲ Polar bears are well suited to the extreme cold of the Arctic region and are skilled at hunting the food sources there.

LOOKING AT AN ENVIRONMENT

This photo montage shows a particular **ecosystem**. There are three different populations in the montage. They are a butterfly, an otter, and a wolf. These three species occupy the same environment because there is enough food, shelter, and breeding ground for all of them.

Together, these three species make a community. In this case, they can be described as a "wildlife community" that occupies the same habitat. The photo montage also shows some of the nonliving factors of this ecosystem. There is a water source for the animals, as well as shade provided by the trees.

CHANGING ECOSYSTEMS

Ecosystems are constantly changing. This can be because populations leave to migrate or because they have been hunted. Indirect human activities are also threatening ecosystems. Humans cut down forests for wood for building, as well as to create space in which to live and farm. Rocks are mined for minerals, and humans also burn fuels that are thought to contribute to global warming, a factor that has a significant influence on the temperature of our climate.

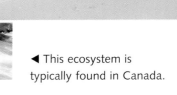

◄ This ecosystem is typically found in Canada.

THE MILLENNIUM ECOSYSTEM ASSESSMENT

Earth's ecosystems are an important source of food, water, and fuel for humans. In 2005, an international study was published, assessing the conditions of Earth's habitats and ecosystems, the potential impacts of any changes, and their ability to meet human needs. The Millennium Ecosystem Assessment used research from satellite images to study changes in coastlines, mountains, wetlands, and urban sprawl over a four-year period. The results will help to direct future initiatives that are used to protect the environment (for more information, see www.millenniumassessment.org).

Adapting to survive

When we talk about a habitat, we mean the environment in which organisms live, as well as the physical factors that affect that environment, such as temperature and rainfall. Most animals have adapted through evolution to suit a particular set of conditions. But if these conditions change, the species must undergo further adaptations in order to ensure its survival.

Physical factors may not be the only change within a habitat. All animals have a particular position within a community. We call this a "niche." If one species becomes better adapted than another, and if they share the same food source or living space, competition can become very fierce.

Some organisms adapt by altering their color for camouflage or growing thick fur during the winter. Others alter their breeding cycles so that young are produced at favorable times of the year—when predator numbers are low or when weather conditions and food supplies are at their least harsh.

▼ Giraffes are unusual-looking creatures, and the reasons behind their characteristics have been widely debated by scientists.

THE GIRAFFE DEBATE

One of the most widely discussed examples of adaptation is the giraffe. Giraffes are herbivores. They eat leaves from trees and live successfully among carnivores without competing for food. Before Darwin published his theory of evolution (see page 6), Jean Baptiste Lamarck (1744–1829) was a popular zoologist who suggested two ways in which organisms could adapt. Lamarck thought that species adapted through use and disuse of behavior and that physical characteristics could be passed on to future generations. According to Lamarck, giraffes once had a neck no longer than a zebra. However, as giraffes began to stretch their necks to eat off tall trees (a steady source of food available to no other animal), they passed this new characteristic to their offspring. However, today, biologists do not all agree with Lamarck's views.

Criticisms of Lamarck's theory include:

The genes that influence neck length cannot be altered during a lifetime by stretching necks.

Long necks didn't evolve solely for feeding at high levels—female giraffes have managed to survive with necks that are a foot (30 cm) shorter than male giraffes; female giraffes have been reported feeding half of their time with their necks horizontal; and both sexes feed much faster when their necks are bent.

If only tall trees were available for food during droughts, then we would expect other animals (such as deer or antelope) to have grown long necks, or to have become extinct.

Other animals have adapted to reach high foliage in other ways. For example, goats are capable of climbing trees to eat high leaves.

▲ In drought conditions, when there is little else to eat, goats will climb up trees to eat the foliage there.

ALTERNATIVE THEORIES

Today, many biologists believe that giraffes accidentally developed long necks through a series of genetic mutations that had nothing to do with use or need. The giraffes with the longest necks found them beneficial because they were able to feed on taller trees. Male giraffes also used their long necks to fight each other for females. Over time, giraffes with shorter necks either died from starvation or lack of nutrition, or their inferior fighting capabilities led to fewer offspring.

DID YOU KNOW?

▶ Eastern and western meadowlarks have been identified as separate species despite the fact that their habitats overlap and they look quite similar. Western meadowlarks (below) occupy a similar space in North America to eastern meadowlarks but avoid competition for mates by having a different birdsong. Since the birds recognize each other through their singing, they do not provide fierce competition for each other.

▶ Baby walruses spend around 16 months in their mother's womb. Sometimes, the embryo does not become implanted in the womb for up to five months to ensure that the baby is born at a more favorable time of the year.

Organisms have adapted to occupy most places on our planet—some of which are extreme in temperature. No animal can carry out its whole life cycle at a temperature higher than 122 °F (50 °C), but plants seem to be more tolerant. At lower temperatures, most animals die, but some are able to tolerate freezing temperatures without the water in their body turning to ice.

SURVIVING THE HEAT

Animals and plants that live in very hot temperatures have adapted to withstand the heat. They may have pale coats to reflect the hot sunshine (like camels) or special mechanisms for preventing water loss, like a cactus (see page 6). A species of algae called *Synechococcus* is found in hot springs where temperatures can be as high as 167 °F (75 °C). Bacteria are even more temperature tolerant, and some have been located in the hot springs of Yellowstone National Park, where temperatures are as high as 198 °F (92 °C).

▲ Bacteria have been found thriving in the almost-boiling hot springs of Yellowstone National Park.

Here are some of the ways in which the body acclimatizes to hot conditions:

Sweat	When we sweat, heat is removed from the body as the liquid evaporates. Some animals have special glands that produce sweat. Others, such as dogs, evaporate heat from parts of their bodies, such as the tongue.

Vasodilation	When we are very hot, we turn red as blood vessels near our skin's surface dilate, and other blood vessels travel to the surface of the skin. When blood travels nearer to the surface, some of its heat is lost through radiation.
Physical changes	Some animals are active at night to avoid the heat. During the day, animals seek shade, and people wear light clothes to stay cool. In hot weather, you can feel thirsty because water has been lost from your body.

SURVIVING THE COLD

Animals living in extreme cold temperatures (such as at the poles) face the greatest challenge of maintaining body warmth. Temperatures here can fall to as low as -22 °F (-30 °C). Arctic animals stay warm in the following ways:

Layer of blubber	This layer of fat found around the animal's skeleton and organs is insulating and helps to keep the body core warm. Arctic animals also tend to be round in shape to keep isolated body parts from freezing.
Remain moving	Some animals move much more in the Arctic than they would in tropical regions. Movement generates heat to keep muscles and organs warm.
Vasoconstriction	Blood vessels change in volume so that they lie much deeper in the skin's surface. Heat is retained in the blood and can then be passed through vital organs to distribute the warmth.
Special features	Some arctic fish contain a special substance in their blood, similar to antifreeze, which allows them to swim at temperatures below freezing.
Physical changes	Arctic animals acquire a thick winter coat to keep them warm in colder conditions. Humans have learned from these animals and sometimes wear animal skins in the winter to keep warm.

Plant life has also adapted to the cold. Plants can't push their roots into frozen soil, and if plant cells freeze, they die. However, many plants lie dormant under a blanket of snow and wait until the summer when the ground thaws to grow and reproduce. Some arctic plants have shallow roots that skim the top of the soil, and lichens have adapted to survive on bare rock. Most arctic plants are dark to absorb as much heat as possible, and they grow in clumps to break up the cold winds and protect each other.

▼ These lizards are basking in the sun to keep themselves warm.

HOT AND COLD BLOOD

The temperature of an animal's blood is related to its body temperature. Warm-blooded creatures try to keep their body temperature at an almost constant level. They generate their own heat by converting the food they eat into energy. Cold-blooded creatures can't control their body temperature, but instead rely on the temperature of their surroundings. They can be active only when they are warm.

DID YOU KNOW?

▶ Penguins live in one of the most extreme conditions inhabited by any warm-blooded animal on Earth. Penguins huddle together in groups of up to several thousand to keep warm when it gets cold. Penguins also avoid contact with the snow and ice by rocking back on their heels and holding their toes up. They support themselves by their stiff tail feathers, which have no blood flow and therefore lose no heat against the snow and ice.

▶ In 1999, tiny bacteria were found two miles (3 km) beneath the seabed in Western Australia. At this depth, temperatures exceed 300 °F (150 °C). The microbes are referred to as nanobacteria because they are so tiny, measuring 20 nanometers (0.0000008 in.) across. When the bacteria were brought to a laboratory for study, they appeared to get much larger, indicating that they were trying to adapt to a cooler climate.

◀ Penguins that live in the coldest climates tend to have longer feathers and thicker blubber. Grouping together also helps them keep warm.

▶ The marine iguana (right) is capable of shrinking by as much as 20 percent when the weather becomes warm and food is scarce. This shrinking includes body fat and tissue, as well as skeletal loss. The marine iguana then grows back to full size when conditions improve. Scientists believe that these changes began to occur when weather conditions brought warm water to the marine iguana's habitat, destroying its source of food.

▲ The marine iguana is well suited to the cold waters of the Galapagos Islands. When food becomes scarce, however, they have been known to shrink by up to 2.75 inches (7 cm) over a 2-year period.

HIGH AND LOW

Species also have to adapt to areas of high and low altitude. At sea level, about 20 percent of the air is made up of oxygen, and there is enough air pressure to inflate our lungs and to breathe efficiently. At higher altitudes, the air pressure falls. At 19,700 feet (6,000 m), the air pressure is not enough to efficiently inflate our lungs.

DID YOU KNOW?

▶ Professional climbers have reported seeing birds flying comfortably at high altitudes, where air pressure is low. These birds achieve this feat because their lungs have adapted so that oxygen from the air is absorbed into the lungs in both inhalation and exhalation. The birds also maintain a good flow of oxygen-filled blood to their brains.

The average person will begin to feel a decrease in physical performance at about 9,800 to 13,100 feet (3,000–4,000 m). Breathing will become more difficult, and altitude sickness can result. If you breathe quickly, you lose more carbon dioxide, and the amount of blood flowing to the brain can be affected. Other symptoms of altitude sickness include headache, nausea, and confusion. It can

take 10 to 14 days to become acclimatized to high altitudes. During this time, climbers take things slowly. Altitude sickness can be fatal in some cases, unless appropriate action is taken.

ADAPTATIONS TO HEIGHT

The highest point at which humans are known to live is found in the Central Andes at mountain heights of 14,000 feet (4,300 m). Some well-trained mountain climbers have successfully reached the summit of Mount Everest at 29,030 feet (8,848 m) without oxygen, but to live at such heights without oxygen is impossible.

Andean people have adapted to live at high altitude.

The lungs of Andean people are 25 percent larger than those of the average human. This means that more oxygen can be taken in with each breath, making breathing more efficient so that enough oxygen is available for the body to use at any given time.

Andean people have more red blood cells in their blood than most people. A larger number of red blood cells means that more oxygen can be carried around the body.

Andean people have a higher volume of blood, making oxygen delivery even more efficient.

A llama is a good example of a mountain animal that has genetically adapted to live at high altitudes. Llamas have adapted to have more hemoglobin—the component that carries oxygen around the body—in their blood.

BENEATH THE EARTH

Humans generally live above sea level (although some countries, such as the Netherlands and Bangladesh are low-lying), but many organisms live underground or in the far depths of the oceans. In cave systems, species have been found that have adapted to a lack of light. Although they don't have any eyes, they navigate their way around using antennae. Some species have adapted to spend their whole life in the dark zone of a cave and would be unable to survive in surface environments. Organisms have also been found at the bottom of the sea. There is a lack of sunlight here, but they get their energy from hot sulfur springs. The springs are heated by magma that spews up through cracks in the ocean floor.

▲ Llamas live in the mountains of South America. They have adapted well to altitudes of around 16,400 feet (5,000 m).

DID YOU KNOW?

▶ Many successful athletes have come from high altitudes. Ethiopian Haile Gebrselassie (below right) lives in Addis Ababa at an altitude of 7,875 feet (2,400 m) and trains in the surrounding hills. Ethiopians can run long distances at altitudes where oxygen pressure is lowered, so they find running at sea level much easier.

Time travel: Human threats

In 1500, the world's population was about 435 million. Today, it stands at about six billion, and a million new children are born each day. This dramatic increase is due primarily to significant improvements in medicines, food production, and living conditions. However, humans are still threatened by disease and other natural disasters.

Disease

Six diseases are responsible for 90 percent of deaths from infectious diseases around the world. These are influenza (flu), HIV/AIDS, diarrheal diseases (such as cholera), tuberculosis, malaria, and measles. People living in densely populated cities are most at risk.

Flu: Although a vaccine is available against influenza, the virus is constantly changing, making the vaccine less effective. There were three major flu epidemics in the 1900s—in 1918, 1957, and 1968—when new flu strains surfaced in humans. The most severe, in 1918, killed at least 25 million people worldwide.

HIV/AIDS: AIDS is a disease of the immune system, caused by the human immunodeficiency virus (HIV). At least 70 percent of all HIV cases are found in the African countries below the Sahara desert, although only a tenth of the world's population lives there. AIDS affects up to 40 percent of the population of some African countries, such as Botswana, and aggressive AIDS awareness programs are now in force, teaching "safe sex" to prevent the disease from spreading and offering milk to breastfeeding mothers who are infected with the virus and could pass the disease on to their child. Free drug therapy is also available to pregnant women. Despite

▲ The virus that causes AIDS (HIV) infects the white blood cells of the immune system.

these efforts, the chances of a 15-year-old boy dying from AIDS is 65 percent in South Africa and 90 percent in Botswana. In the Western world, there is more money available for medication. In New York City, for example, an annual AIDS walk raises approximately $4.5 million to help people living with HIV and AIDS.

▲ Many parts of the developing world do not have access to clean water for drinking or washing, and this can lead to serious diseases.

Cholera: About 1.1 billion people in the world live without clean water and are at high risk of contracting water-borne diseases, such as cholera. Another 2.4 billion people are also at risk because they do not have adequate sanitation facilities. Cholera is a serious infection of the intestines caused by eating food or drinking water that has been contaminated with the bacterium *Vibrio cholerae*. Symptoms include severe diarrhea that can quickly lead to dehydration and death if treatment is not promptly given. Cholera kills about 5,000 people every year and remains a serious threat in developing countries where access to safe drinking water and adequate sanitation cannot be guaranteed.

Tuberculosis (TB): About one-third of the people in the world are carriers of TB. As more people travel around the globe, the disease is becoming more widespread. TB is a bacterial disease that is spread through the air by coughs and sneezes. Those living in close proximity to others are most at risk of the disease spreading. TB can be treated with antibiotics, but the bacteria are becoming more resistant to these drugs. TB is often linked to poor countries, where people live in squalid conditions, but the disease is making a comeback in the developed world, largely due to vulnerable populations, such as street people and drug users, whose immune systems may not resist the disease. Most people who "carry" tuberculosis, however, remain disease-free.

Malaria: Malaria occurs in more than 100 countries worldwide, and more than 40 percent of the people in the world are at risk from the disease because they live or travel in these regions. Scientists also fear that numbers will increase if Earth's temperatures rise due to global warming. Mosquitoes carry the disease, and the warmer weather allows mosquitoes to breed much faster and therefore bite more often. Malaria kills about a million people in the world every year, and most of these are children. More than 90 percent of these deaths are in Africa. Malaria is estimated to cost Africa $12 billion dollars a year.

▲ Malaria is a disease of the blood that is carried by mosquitoes.

▲ Diseases like tuberculosis are more likely to spread in squalid conditions, where people live in close proximity without adequate sanitation facilities.

Measles: Measles is a highly contagious viral disease that causes death in extreme cases. A vaccine became available in 1963, but measles still kills about 900,000 people a year and infects a total of 30 million. Measles is the world's most preventable disease in children, but many still die from it. This is largely because governments do not have the money to have a national vaccination program, and communities that struggle to find food on a daily basis cannot fund one themselves.

FAMINE AND DROUGHT

Severe droughts are common in Africa and can lead to mass starvation when crops fail due to lack of rain. However, in many cases, poverty is the most significant cause of food scarcity. The cultivation and exportation of cash crops to make money at the expense of everyday foods is also a problem. Aid agencies try to alleviate these conditions, but political circumstances—such as local wars, tribal clashes, and a lack of effective government—complicate matters.

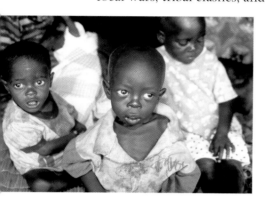

◀ The World Health Organization estimates that a third of the world's population is starving. Every year, 15 million children die of hunger.

TIME TRAVEL: INTO THE FUTURE

▶ Scientists are beginning to speculate that in just a few generations, the spread of AIDS could alter human evolution in some African countries. In the past, infectious diseases, such as the bubonic plague and measles, have changed human genetic evolution because only the fittest survived beyond their reproductive years. This genetic resistance does not come without its downfalls, however. In the 1940s, a link was found between malaria and sickle-cell anemia, a serious hereditary blood disease in which the number of oxygen-carrying red blood cells is reduced. Scientists discovered that people with sickle-cell disease had a greater resistance to malaria. They found that a genetic mutation that occurred naturally in the fight against malaria led to this serious blood condition.

WAR

Wars kill millions of people and can affect anyone around the world. In the 1900s, wars were responsible for huge numbers of deaths, as the table below shows.

Event	Number of deaths*
World War I (1914–1918)	7.1 million
World War II (1939–1945)	39.8 million
Korean War (1950–1953)	1.7 million
Vietnam War (1965–1975)	1.5 million
Iran/Iraq War (1980–1988)	1 million

* Estimates are difficult to make because statistic sources vary.

DEATH

A million new children are born every day. Thanks to advances in medical science, diet, and improved living conditions, deaths from natural causes are falling. This means that as more people are born than die, the world's population is growing. Life expectancy increased dramatically in the 1900s, particularly in developed nations. In 1901, average life expectancy at birth in the U.S. and the UK was around 49 years. Today, it is around 77 years. However, in recent years, Africa, Asia, and South America have seen a reduction in life expectancy due to the onset of AIDS. Africa is now home to the lowest life expectancy—around 37 years.

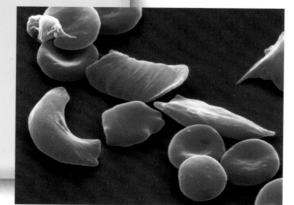

◀ This microscopic image compares normal red blood cells (rounded) with distorted sickle-shaped cells (pink). Sickle-cell anemia is a blood disease in which some red blood cells do not carry enough oxgyen. Magnification 3,000x.

Ecosystem change and survival

Over millions of years, planet Earth has undergone some dramatic changes. As Earth has evolved, a few catastrophic events—from ice ages to meteorite collisions—have threatened species or wiped them out completely.

THE ICE AGES

An ice age is a period of time during which large areas of the globe are covered in ice sheets that form when Earth's glaciers spread out over the warmer regions of the planet.

SURVIVING AN ICE AGE

When an ice age begins, plants are the first species to suffer. Few plants can survive an ice age, and with a significant loss of producers in any food chain, herbivores have to adapt to another food source or migrate to milder regions. If herbivore numbers decline, only those carnivores that are best at hunting prey are likely to survive.

The last ice age was about 10,500 years ago and was responsible for the extinction of saber-toothed cats, giant sloths, woolly rhinos, and other large mammals.

Ice ages can be caused by:

Changes to Earth's orbit—The major cause of ice ages is believed to be a shift in the tilt of Earth's axis in orbit. This can alter the amount of the sun's energy reaching Earth and cause a reduction in temperature.

Continents changing positions—Ice ages occur when large areas of land are positioned in the north or south of the globe, causing temperatures to drop and extensive glaciers to form.

Continents lifting—When Earth's surface plates collide, they can cause mountains to form. These changes in Earth's surface cause snow to fall, which reflects the sun's heat and cools Earth. Glaciers also form at these high altitudes and soon persist at lower altitudes as temperatures drop.

A reduction of carbon dioxide—Carbon dioxide in Earth's atmosphere keeps the planet warm enough for life to exist. Without this blanket, Earth would become icy and too cold for plants and animals to survive.

▶ Saber-toothed cats became extinct when the large prey they fed on retreated with the glaciers during an ice age.

METEORITES

The most notable mass extinction was the downfall of the dinosaurs around 65 million years ago. Scientists believe that a huge meteorite from outer space hit Earth at this time, causing fires and large clouds of dust that blocked the sun. Such a dramatic collision would have made Earth very cold and dark, so dinosaurs and other life could not survive. Scientists think that this meteorite made a huge crater found in the Yucatán area of Mexico.

It is estimated that the meteorite collision killed about 75 percent of the world's species. Dominant reptiles on land, sea, and air perished. Only small creatures such as mammals, birds, and fish would have been able to survive in such harsh conditions. Over time, these creatures grew and evolved to give us the wide range of wildlife we see today.

Some of the most significant extinctions are summarized below:

Time (years ago)	Possible cause*	What was made extinct?
440 million	Falling and rising sea levels as glaciers formed and then melted.	85 percent of all species.
380 million	Unknown.	82 percent of species, including coral reef invertebrates.
245 million	Comet/meteorite impact or lava flood.	96 percent of all species, including sea scorpions.
199–214 million	Massive floods of lava erupting.	74 percent of marine species.
65 million	Meteorite impact.	76 percent of species, including dinosaurs.

* Causes are still debated.

▶ Dinosaurs used to rule Earth but became extinct 65 million years ago. The causes are still debated.

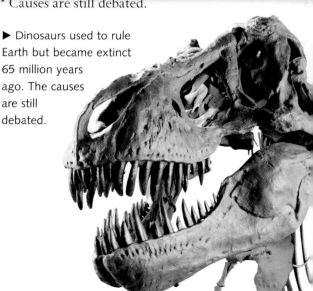

▲ An artist's impression of the meteorite impact that may have caused the extinction of more than 70 percent of species 65 million years ago.

▲ Volcanoes occur where Earth's crust is weak, and boiling-hot molten rock erupts from beneath the surface.

VOLCANOES

Fortunately, we are unlikely to see an ice age or meteorite collision in our lifetime, but there are more significant natural impacts that affect our everyday lives as well as the living things that we share our planet with.

Volcanoes usually occur in mountainous regions where boiling-hot liquid rock (called magma) escapes from the center of Earth and breaks through a thin crust of Earth's surface.

Volcanoes can cause:
▶ A release of toxic gases
▶ A fast flow of magma
▶ Large volumes of ash

In 1926, a village of 300 people perished beneath Bali's Mount Batur when a volcano warning was

▲ Active volcanoes are extremely dangerous, but people still choose to live nearby because the soil is rich and fertile and ideal for farming.

given and they refused to leave their homes. Boiling-hot magma burns any plant or animal life that falls in its path in an instant.

It can take a long time for Earth to recover from a volcanic eruption. The ground remains very hot—sometimes several years later. However, plant life does begin to recover, and with the arrival of lichens and bamboo, many food chains reestablish themselves. Many people choose to live in the foothills of volcanoes because the soil there is full of nutrients, such as iron, that have been carried with the magma from Earth's core.

FOREST FIRES

Forest fires are likely to occur in regions where the weather is very hot and dry. They can be caused by lightning or by human activities, such as campfires or sparks from machinery. Once a forest fire begins, it quickly spreads, destroying all life in its path. Plants or animals in the forests do not have the ability to escape, and with a loss of vegetation, many food chains take a long time to recover. In 2003, a forest fire in Colorado destroyed 12 homes and 3,500 acres (1,400 ha) of forestland, and a week later, small patches of smoke were still seen from the burning embers.

Fire has a devastating effect, but scientists now believe that fire is actually an essential part of the maintenance of a forest ecosystem. In the aftermath of a forest fire, some plants begin to grow again, while other new species arise. Many animal species also take advantage of the new habitat that arises and the different types of growing vegetation.

LANDSLIDES

Landslides occur when large parts of land move down a slope. They can be caused by atmospheric changes, a movement of Earth's crust, human activities, or a combination of these things. Avalanches are landslides that occur in snow. They contain about 110,000 tons (100,000 t) of snow and are capable of burying people and wildlife in seconds.

FLOODING

Floods are high levels of water that occur above the banks of a river or stream. Floods usually occur because of prolonged and unexpected rain or when a river bursts its banks. Wildlife can drown or lose its natural habitat, and if floodwater lingers for long periods of time, plant life will rot and die. Dams are built in high-risk areas to hold back floodwaters and preserve natural habitats.

▲ In 2002, forest fires destroyed 9,650 square miles (25,000 sq km) of forest and wildlife habitat.

TEST YOURSELF

▶ Can you think of any other types of natural disasters that affect habitats?

HUMAN ACTIONS

Ecosystems are constantly changing due to the forces of nature, but the actions of humans play a part, too. Our increased reliance on fuels for cars and our need for extra living space are having catastrophic effects on weather patterns and the wildlife of our planet.

GLOBAL WARMING

Earth is surrounded by a blanket of atmospheric gases, which we call "greenhouse gases." This blanket is vital because it makes Earth warm enough for species to survive. However, when we burn fossil fuels (such as coal, natural gas, and oil) to heat our homes, cook our food, and run our vehicles, carbon dioxide is released into the atmosphere, and the blanket of gases gets thicker. Unfortunately, although the sun's radiation can reach Earth, when it bounces off Earth's surface, a change in wavelength means that the sun's radiation cannot escape through this thicker blanket, and Earth's temperature rises. Over the last 100 years, Earth's average temperature has increased by 1.8 °F (1 °C), and scientists believe that it is set to rise by another 3 °F (1.7 °C) by the year 2050. Carbon dioxide is one of the main culprits of global warming. However, methane gas from rotting plants or animal dung is another contributor. The effects of global warming include:

▼ Global warming occurs when an increase in greenhouse gases causes some of the sun's radiation to become trapped in Earth's atmosphere.

Blanket of "greenhouse gases"

Some energy is trapped by Earth's atmosphere

Some energy escapes from Earth's atmosphere

Energy from the sun

Thinning polar ice caps and shrinking glaciers. Those species that rely on polar ice can quickly find themselves homeless. Polar bears, seals, and penguins are all at risk.

Rising sea levels caused by the melting of ice caps or the expansion of water as it heats. This would have a drastic effect on species living in coastal areas. It might take centuries for the ice caps to melt, but the effects would be catastrophic—the ice covering Greenland would melt, and if sea levels rose by 23 feet (7 m), London would completely disappear.

A change in reproductive cycles. Scientists know that some European plants, for example, are flowering a year earlier than they used to because of rising temperatures. Birds and frogs have also been found to breed earlier in the season.

A loss of plant and animals species. Because global warming is occurring at a fast rate, scientists are worried that species will not have time to adapt. It is estimated that at least 1 million species (18 percent of all land plants and animals) will be on their way to extinction by 2050.

A loss of coral reefs. Coral reefs are home to many plants (called algae) and small organisms. As sea temperatures rise, many coral species die out because they can't adapt to the warmer waters.

INTERNATIONAL EFFORTS

The Kyoto agreement is an international set of rules established in 1997 to curb the effects of global warming. To date, 141 countries have agreed to reduce their emissions of greenhouse gases by 29 percent by the year 2010. Notable exceptions include Australia and the U.S. (who feel that the changes would be too costly to introduce at the moment) and large developing countries such as India, China, and Brazil (who are not required to meet specific targets yet). The problem is that the agreement requires a reduction in the use of fossil fuels—currently the main source of energy for most countries. Countries that have signed the agreement are putting projects in place to reduce carbon dioxide emissions in industry (including the introduction of taxes to encourage action) and to conserve forest areas that help to remove carbon dioxide from the atmosphere. In the UK, there are currently concerns that greenhouse gas emissions caused by air travel will make it very difficult to meet the agreed targets. Air travel in the UK has doubled greenhouse gas emissions from aviation in the last 13 years. Aviation emissions in the UK rose from 22 million tons (20 million t) in 1990 to 43.5 million tons (39.5 million t) in 2004.

▲ A flight from the UK to Australia produces 7,900 pounds (3,600 kg) of carbon dioxide per person.

DEFORESTATION

Trees are vital for the balance of Earth's atmosphere because they remove carbon dioxide during the process of photosynthesis. However, humans are destroying large areas of forest on a daily basis to create:

▶ Wood for fuel or building
▶ Space for farming and living
▶ Space to grow other crops
▶ Access for mining valuable resources

Developed nations removed most of their forests a long time ago, but now the focus is on the developing world, where deforestation is becoming commonplace. About one percent of the world's forests are destroyed every year, and biologists predict that the rain forest in Brazil will be completely destroyed in about 40 years, destroying more than 30 percent of the world's animal and plant species at the same time.

WHAT IS BEING DONE?

In some risk areas, farmers are being taught how to plant crops that stimulate forest growth. In the foothills of Kilimanjaro in Africa, for example, farmers are planting crops, such as coffee, near their homes and avocado trees farther up the slopes. This progression of plants provides protection for some wildlife.

▲ Trees are often planted where coffee is grown to provide shade for the crop. In Puerto Rico, many forests are now developing where coffee was once grown.

In developing countries, woodland areas are also managed to preserve habitats. Coppicing is an ancient form of woodland management that involves cutting trees to ground level in the winter months and leaving them to grow. When trees reach a harvestable size, they are cut again.

This "cycle" means that wood can be taken without destroying forest life. Felling and thinning is also used—selected wood supplies are cut down to create space and to encourage the growth of younger plants.

▲ Coppicing is the art of cutting trees and shrubs to ground level to encourage regrowth and a sustainable supply of timber for future generations.

Glossary

ADAPT – To adjust to a different situation.

ANDEAN – Describes a person who lives in the Andes mountains of South America.

BIOACCUMULATION – The accumulation of a substance, such as a toxic chemical, in various tissues of a living organism.

BIOMAGNIFICATION – The increase in concentration of a substance, such as a toxic chemical, as it moves up the food chain.

CAMOUFLAGED – Disguised. Camouflage is the method by which an otherwise visible organism blends in with its surroundings. A tiger's stripes act as camouflage, for example.

DORMANT – A period of inactivity. Many plants and animals lie dormant during the winter months.

ECOSYSTEM – A collection of living things and the environment in which they live.

EVOLUTION – A change in the traits of living organisms over generations.

EXTINCT – To no longer exist or live. The moment of extinction is generally considered to be the death of the last individual of a species. It is estimated that more than 99.9 percent of all species that have ever lived are now extinct.

FERTILIZE – To join together male and female reproductive cells to start the growth of a new organism. The ova (egg cells) in flowers are fertilized by the male pollen cells. The egg cells in animals are fertilized by the male sperm cells.

GENETIC – Having to do with genes, the hereditary units containing the code for particular characteristics. The study of genes is called genetics.

GLOBAL WARMING – A sustained increase in the average temperature of Earth's atmosphere, which can cause climate change.

ANSWERS

page 7 Test yourself
Camels: double eyelids for protection against sandstorms; wide feet to prevent sinking into the sand; a hump to contain fat that serves as a source of water; pale color to help with temperature control; nostrils that can open and close to prevent sand from going up the nose; long eyelashes to keep sand out of the eyes; can withstand a high body temperature without the need to sweat; thick lips to eat the prickly desert plants without feeling pain. Migratory birds: flying south to a place where there is likely to be more food due to better weather conditions helps to ensure survival.

page 9 Investigate
Loris: enormous eyes (aids nighttime vision); big ears (can hear predators coming from a distance);

grasping fingers (lives in trees, and this helps with balance).
Skunks: smelly spray (repels most predators); black and white color (can be difficult to see in the dark).

page 13 Test yourself
Tiger: large and fast to run; camouflaged to get near prey; excellent hearing and sight.
Hare: large ears to hear predators; excellent eyesight—eyes are on the sides of the head to give more all-around vision; an extremely fast sprinter.

page 16 Test yourself
(1) Consumers use energy for growth, respiration, and reproduction; energy is also lost in excretion.
(2) Herbivore—eats plants; omnivore—eats plants and animals; carnivore—eats animals; insectivore—eats insects.

(3) Example answer—grass, cow, human.
(4) Producer—grass and plants; primary consumer—wildebeests; secondary consumer—lion.

page 24 Investigate
The last pyramid appears to be upside-down.

page 42 Test yourself
Example answers—earthquake, tsunami, tropical cyclone, tornado.

page 45 Investigate
(1) Reduces the need for heating.
(2) Reduces the need for heating.
(3) Reduces fuel consumption.
(4) Reduces fuel consumption.

HABITAT – The area where an organism normally lives.

HIBERNATION – The practice of spending part of the cold season in a dormant state as protection against cold conditions or scarce food resources.

HORMONES – Chemical substances produced by the body by glands that are transported by the blood to other organs to stimulate their function. Hormones are chemical messengers.

MAGNETIC FIELD – An area of force that can be exerted on a magnet. Earth has a magnetic field, extending between the North and South Poles, which is caused by the movement of electrical charges.

MIGRATED – Moved from one place to another. Many animals migrate to avoid cold conditions or a shortage of food.

NATURAL SELECTION – A theory proposed by Charles Darwin that only the organisms which are best suited to an environment survive and pass their genes on to future generations.

NOCTURNAL – Active at night.

NUTRIENTS – Substances that are nourishing or provide food for a living organism.

ORGANISMS – Living plants or animals.

PHOTOSYNTHESIS – The process in green plants in which foods (mainly sugars) are made from carbon dioxide and water, using energy from sunlight.

POACHED – Taken illegally.

POLLINATION – The transfer of pollen grains from the male anther to the female stigma in flowering plants.

POLLUTION – Contamination of the environment as a result of human activities.

PREDATORS – Organisms that live by preying on other organisms.

PREY – An animal that is hunted for food.

PRIMATE – A group of mammals that includes monkeys and apes. Primates have limbs, and hands and feet capable of grasping.

REPRODUCE – To produce offspring.

RESISTANCE – The ability of an organism to defend itself against a disease or harmful agent.

SELECTIVE BREEDING – The process of developing a particular breed of animal over time using strict reproduction techniques.

SPECIES – A plant or animal type.

STRAIN – A different variety of the same species. Different strains of the flu virus can develop, for example.

TOXIC – Capable of causing injury or death by poison.

TROPHIC – The food relationship of different organisms within a food chain. Each part of a pyramid of numbers is called a trophic level.

Index

Page references in italics represent pictures.

PHOTO CREDITS – *(abbv: r, right, l, left, t, top, m, middle, b, bottom)* **Cover background image** www.istockphoto.com/Jennifer Trenchard **Front cover images** (tr) Scott Camazine/K. Visscher/Science Photo Library (lm) www.istockphoto.com/Christina Craft **Back cover image** (inset) www.istockphoto.com/Christina Craft **p.1** (tr) www.istockphoto.com/Gary Forsyth (bl) www.istockphoto.com/Kevin Russ (br) www.istockphoto.com/Andrew Caballero-Reynolds **p.2** (bl) wwwistockphoto.com/Jeffrey McDonald **p.3** (tr) www.istockphoto.com/Brian Palmer (br) www.istockphoto.com/Andy Didyk **p.4** (tl) www.istockphoto.com (mr) David Haring/Oxford Scientific (bl) Joe Tucciarone /Science Photo Library **p.5** Mark Jones/Oxford Scientific **p.6** (tr) Stan Osolinski/Oxford Scientific (br) Bettmann /Corbis **p.7** (mr) David Fox/Oxford Scientific **p.8** (bl) www.istockphoto.com/Andrei Botezatu (tr) Richard Herrmann/Oxford Scientific **p.9** (mr) www.istockphoto.com/Malcolm Romain (mr) David Haring/Oxford Scientific **p.10** (bl) www.istockphoto.com/Kevin Russ (br) www.istockphoto.com/Alexandre Caron **p.11** (tr) Daniel Cox/Oxford Scientific **p.12** (bl) www.istockphoto.com/Steve Geer (mr) Steve Turner/Oxford Scientific **p.13** (ml) Survival Anglia/Oxford Scientific **p.14** (ml) Christian Jegou, Publiphoto Diffusion/Science Photo Library **p.15** (ml) www.istockphoto.com/Andre Maritz (br) www.istockphoto.com/Andy Didyk **p.16** (mr) Mary Beth Angelo/Science Photo Library **p.18** (ml) Nicole Duplaix/Corbis (br) Survival Anglia/Oxford Scientific **p.19** (bl) Brooks Kraft/Corbis **p.20** (tr) Reuters/Corbis (br) Edward Parker/Oxford Scientific **p.21** (br) Annie Griffiths Belt/Corbis **p.22** (bl) Michael Fogden/Oxford Scientific **p.23** (ml) Peter Menzel/Science Photo Library (mr) Alan Sirulnikoff/Science Photo Library **p.25** (mt) www.istockphoto.com/Laurie Knight (mmt) www.istockphoto.com/ Keith Naylor (mmb) www.istockphoto.com/Thomas Mounsey (mb) www.istockphoto.com/Zibi Kolo (bt) www.istockphoto.com/Gary Forsyth (bm) www.istockphoto.com/Jeffrey McDonald (bb) www.istockphoto.com **p.26** (t) NOAA (mr) Geray Sweeney/Corbis **p.27** (br) Daniel Cox/Oxford Scientific **p.28** (tl) www.istockphoto.com/Daryl Faust (m) The Travel Library Limited/Oxford Scientific (mr) www.istockphoto.com/Tom Lewis (bl) www.istockphoto.com /Richard Foote **p.29** www.istockphoto.com/Jane Norton **p.30** (tr) John and Lisa Merrill/Corbis (bl) Darrell Gulin/Corbis (br) Carleton Ray/Science Photo Library **p.31** (tr) www.istockphoto.com/Jason Pachero (bl) www.istockphoto.com/Geoff Kuchera (br) Norman Reid **p.32** (mr) Daniel Cox/Oxford Scientific (r) Doug Allan/Science Photo Library **p.33** (l) Daniel Cox/Oxford Scientific **p.34** (l) Mark Jones/Oxford Scientific (br) Danny Lehman/Corbis **p.35** (l) R. Sponlein/zefa/Corbis (br) Dan Heims/Newsport/Corbis **p.36** (tr) Annie Griffiths Belt/Corbis (bl) BSIP, Cavallini James/Science Photo Library **p.37** (l) Johnny Buzzerlo/Corbis (r) Sinclair Stammers/Science Photo Library **p.38** (ml) Vanessa Vick/Science Photo Library (bm) Eye of Science/Science Photo Library **p.39** (br) Joe Tucciarone/Science Photo Library **p.40** (bl) D.Van Ravenswaay/Science Photo Library (br) Louie Psihoyos/Corbis **p.41** (t) Kevin Schafer/Corbis (mr) Paul A. Souders/Corbis **p.42** www.istockphoto.com/Andrew Caballero-Reynolds **p.43** (mr) NASA (br) www.istockphoto.com/David Keith Gibbons **p.44** (tr) Reuters/Corbis **p.45** (tl) Tony Arruza/Corbis (br) Frank Lane Picture Agency/Corbis.